SECRET

```
┌─────────────────────────────┐
│                             │
│   SPECIAL OPERATIONS        │
│   FIELD MANUAL –            │
│                             │
│   STRATEGIC SERVICES        │
│        (Provisional)        │
│                             │
└─────────────────────────────┘
```

Prepared under direction of
The Director of Strategic Services

SECRET

SECRET

SPECIAL OPERATIONS FIELD MANUAL
– STRATEGIC SERVICES
(Provisional)

Strategic Services Field Manual No. 4.

SECRET

SECRET

Office of Strategic Services
Washington, D. C.
23 February 1944

This Special Operations Field Manual — Strategic Services (Provisional) is published for the information and guidance of all concerned and will be used as the basic doctrine for Strategic Services training for such subjects.

It should be carefully noted that Special Operations as defined in this Manual covers the following subjects: (1) sabotage; (2) direct contact with and support of underground resistance groups; (3) conduct of special operations not assigned to other Government agencies and not under direct control of theater or area commanders. Special Operations do not include promotion of, or engagement in, guerrilla activities or subversive maritime activities, which will be the subjects of other provisional basic field manuals.

The contents of this Manual should be carefully controlled and should not be allowed to come into unauthorized hands.

AR 380-5, pertaining to handling of secret documents, will be complied with in the handling of this Manual.

William J. Donovan

William J. Donovan.
Director

TABLE OF CONTENTS

SECTION I — PRINCIPLES AND METHODS

1. THE MISSION, OBJECTIVE, AND IMPLEMENTS 1
2. DEFINITIONS 1
3. METHODS 3

SECTION II — ORGANIZATION

4. ORGANIZATION IN THE UNITED STATES . 3
5. ORGANIZATION AT FIELD BASES . . . 3
6. ORGANIZATIONAL FUNCTIONS 4
7. ORGANIZATION OF FIELD OPERATIVES . 4
8. CONTACT WITH AND SUPPORT OF UNDERGROUND RESISTANCE GROUPS . . . 4

SECTION III — PROCUREMENT OF PERSONNEL

9. ORGANIZATION FOR PROCUREMENT . . 4
10. SOURCES FROM WHICH PERSONNEL ARE DRAWN 5
11. TYPES OF PERSONNEL REQUIRED . . . 5

SECTION IV — TRAINING

12. ORGANIZATION FOR TRAINING 5
13. SCOPE OF TRAINING 6
14. TRAINING OBJECTIVES 6

SECTION V — SUPPLY AND COMMUNICATION

15. THE SPECIAL OPERATIONS SUPPLY PROBLEM 7
16. ORGANIZATION FOR SUPPLY . . . 7
17. SUPPLIES FOR SABOTEURS AND RESISTANCE GROUPS 8
18. PAYMENT AND SUBSIDIES 8
19. RADIO AND OTHER SIGNAL EQUIPMENT . 8
20. TRANSPORTATION 8

SECTION VI — COORDINATION OF SPECIAL OPERATIONS ACTIVITIES WITH THAT OF OTHER OSS BRANCHES AND THE ARMED FORCES AND OTHER AGENCIES OF THE UNITED NATIONS

21. COOPERATION WITH OTHER OSS BRANCHES 9
22. COOPERATION WITH SIMILAR AGENCIES OF ALLIED GOVERNMENTS . . . 10

23.	COOPERATION WITH THE ARMED FORCES	10
24.	COOPERATION WITH GOVERNMENT AGENCIES	10

SECTION VII — PLANS AND ORDERS

25.	IMPORTANCE OF PLANS AND ORDERS	11
26.	ORGANIZATION AND RESPONSIBILITY FOR THE PREPARATION OF PLANS AND ORDERS	11
27.	PROCEDURE IN OPERATIONAL PLANNING	11

SECTION VIII — SABOTAGE TECHNIQUES

28.	DEFINITION	12
29.	PLANNING SABOTAGE	12
30.	TRAINING OF SABOTEURS	12
31.	TYPES OF SABOTAGE	13
32.	METHODS OF SABOTAGE	14

SECTION IX — MISCELLANEOUS SPECIAL OPERATIONS FUNCTIONS

33.	ADDITIONAL FUNCTIONS	15
34.	MORALE OPERATIONS ACTIVITY	15
35.	INTELLIGENCE ACTIVITY	15
36.	ASSISTANCE TO THE ARMED FORCES	15
37.	DIRECT CONTACT WITH AND SUPPORT OF UNDERGROUND RESISTANCE GROUPS	15
38.	SPECIAL OPERATIONS NOT ASSIGNED TO OTHER GOVERNMENTAL AGENCIES AND NOT UNDER THE DIRECT CONTROL OF THEATER OR AREA COMMANDERS	16

SECTION X — THE SELECTION OF SPECIAL OPERATIONS TASKS AND MISSIONS

39.	TYPES OF TASKS	16
40.	SELECTION OF MISSIONS TO COORDINATE WITH THE MILITARY PLAN	17
41.	AUTHORIZED MISSIONS	19
42.	TASKS SHOULD BE PRACTICAL	19
43.	MISSIONS MUST BE APPROVED BY THE MILITARY COMMANDER	19
44.	CHECK LIST	19

APPENDIX "A" TO SPECIAL OPERATIONS FIELD MANUAL

CHECK LIST — FOR SO (WASHINGTON)	20
CHECK LIST — FOR SO (THEATER)	24

SPECIAL OPERATIONS FIELD MANUAL
STRATEGIC SERVICES
(Provisional)

SECTION I — PRINCIPLES AND METHODS

1. *THE MISSION, OBJECTIVE, AND IMPLEMENTS*

The mission of the OSS is to plan and operate special services, (including secret intelligence, research and analysis, and morale and physical subversion) to lower the enemy's will and capacity to resist, carried on in support of military operations and in furtherance of the war effort. The mission of the Special Operations Branch is to carry out that part of the OSS mission which can be accomplished by certain physical subversive methods as contrasted with the operations of the Morale Operations, the Operational Groups, and the Maritime Unit. The primary objective of the Special Operations Branch is the destruction of enemy personnel, materiel, and installations.

2. *DEFINITIONS*

a. OVER-ALL PROGRAM FOR STRATEGIC SERVICES ACTIVITIES—a collection of objectives, in order of priority (importance) within a theater or area.

b. OBJECTIVE—a main or controlled goal for accomplishment within a theater or area by Strategic Services as set forth in an Over-All Program.

c. SPECIAL PROGRAM FOR STRATEGIC SERVICES ACTIVITIES—a statement setting forth the detailed missions assigned to one or more Strategic Services branches, designed to accomplish a given objective, together with a summary of the situation and the general methods of accomplishment of the assigned missions.

d. MISSION—a statement of purpose set forth in a special program for the accomplishment of a given objective.

SECRET

e. OPERATIONAL PLAN—an amplification or elaboration of a special program, containing the details and means of carrying out the specified activities.

f. TASK—a detailed operation, usually planned in the field, which contributes toward the accomplishment of a mission.

g. TARGET—a place, establishment, group, or individual toward which activities or operations are directed.

h. THE FIELD—all areas outside of the Western Hemisphere in which Strategic Services activities take place.

i. FIELD BASE—an OSS headquarters in the field, designated by the name of the city in which it is established, e.g., OSS FIELD BASE, Cairo.

j. ADVANCED OR SUB-BASE—an additional base established by and responsible to an OSS field base.

k. OPERATIVE—an individual employed by and responsible to the OSS and assigned under special programs to field activity.

l. AGENT—an individual recruited in the field who is employed and directed by an OSS operative or by a field or sub-base.

m. COVER—an open status, assumed or bona fide, which serves to conceal the secret activities of an operative or agent.

n. CUTOUT—a person who forms a communicating link between two individuals, for security purposes.

o. OPERATIONAL GROUPS—a small, uniformed party of specially qualified soldiers, organized, trained, and equipped to accomplish the specific missions set forth below.

p. RESISTANCE GROUPS — individuals associated together in enemy-held territory to injure the enemy by any or all means short of military operations, e.g., by sabotage, espionage, non-cooperation.

q. GUERRILLAS—an organized band of individuals in enemy-held territory, indefinite as to number, which conducts against the enemy irregular operations including those of a military or quasi-military nature.

3. *METHODS*

The methods to be used by Special Operations are all measures needed to destroy enemy personnel, materiel, installations, and his will to resist. The major classifications of SO methods are;

a. Sabotage.

b. Direct contact with and support of underground resistance groups.

c. Special operations not assigned to other governmental agencies and not under direct control of theater or area commanders.

SECTION II — ORGANIZATION

4. *ORGANIZATION IN THE UNITED STATES*

The Special Operations Branch is included under Strategic Services Operations and is responsible for the following:

a. Sabotage.

b. Direct contact with and support of underground resistance groups.

c. Conduct of special operations not assigned to other governmental agencies and not under direct control of theater or area commanders.

d. Organization, equipment, and training of such individuals or organizations as may be required for operations not assigned to other governmental agencies.

5. *ORGANIZATION AT FIELD BASES*

Each field base will normally include an SO section, the head of which is responsible to the Strategic Services Officer in theaters or to the Chief of OSS Mission in neutral countries, and which will participate in the planning and execution of SO activities in that theater or area. SO personnel both at the base and in the field will be

SECRET

responsible for carrying out the approved SO special programs and such additional operations as may be authorized by the theater commander for that theater or area.

6. *ORGANIZATIONAL FUNCTIONS*

<u>a</u>. At headquarters in Washington and in the theaters of operation the SO units, assisted by other OSS units, are responsible for:

Recruiting	Training
Planning	Supply
Administration	Liaison
Staff work	

<u>b</u>. SO in its activities will be assisted by the intelligence branches, the operating branches, Services and Communications Branches, Schools and Training Branch, Field Photographic Branch, and other OSS organizations.

7. *ORGANIZATION OF FIELD OPERATIVES*

Field operatives work individually or in groups as required by the mission and objective. Many operatives working with the underground must of necessity operate alone. Carefully selected and trained units will be organized specially for specific coup de main projects.

8. *CONTACT WITH AND SUPPORT OF UNDERGROUND RESISTANCE GROUPS*

SO operatives may assist and train agents for contact with and support of resistance groups. In order to perform this function effectively, they must ascertain the needs of the resistance groups, arrange for communications with the base and assist in the delivery of such supplies as can be obtained. On occasion it may be practical for SO operatives personally to serve as leaders of already organized resistance groups.

SECTION III — PROCUREMENT OF PERSONNEL

9. *ORGANIZATION FOR PROCUREMENT*

The SO Branch is charged with the responsibility for procurement of its personnel. Civilian clerical personnel

SECRET

is procured through the Services Branch, both in the United States and abroad. Other personnel, including military and naval, is procured in the United States through the Personnel Procurement Branch and at foreign bases through the Services Branch. At all times military and naval personnel must come within the approved allotment of grades and ratings for the theater set by Washington Headquarters.

10. *SOURCES FROM WHICH PERSONNEL ARE DRAWN*

SO may recruit civilians of United States or other nationalities. By agreement with the armed forces, members of the United States Army, Navy, and Marine Corps may be assigned to OSS and detailed to SO for service. Members of the armed forces of our Allies may be attached to OSS and detailed to SO for duty, in each case by agreement with the authorities of the nation concerned.

11. *TYPES OF PERSONNEL REQUIRED*

a. Base personnel will be either military or civilian and are individually selected for their ability to perform special functions.

b. SO agents and operatives are selected for their intelligence, courage, and natural resourcefulness in dealing with resistance groups. In addition they must have stamina to be able to live and move about undetected in their area of operation. Normally, they should be fluent in the local language and be a native of a nationality acceptable to the authorities and people of the area.

SECTION IV — TRAINING

12. *ORGANIZATION FOR TRAINING*

Basic training courses are provided by the Schools and Training Branch. The Special Operations Branch collaborates with that Branch by developing satisfactory training courses for the schools. Training is a continuous process and it is the responsibility of each SO chief, both in the United States and in the field to see that training progresses satisfactorily.

13. *SCOPE OF TRAINING*

Because of the hazardous nature and specialized technical requirements of SO, it is important that every individual in the organization receive a thorough schooling in the work he has to perform. For field operatives and all those having to do with planning, servicing, and commanding field operatives, training starts with the basic school courses which include instruction in secret intelligence and morale operations as well as special operations. Special schooling for each mission is given to the individuals assigned to it. For specific tasks schooling becomes intensive and detailed and concludes in a final briefing or instruction just prior to the execution of the task.

14. *TRAINING OBJECTIVES*

a. For Operating Techniques

The SO operative must be able to assume perfect cover or concealment. He must know how to employ underground methods of communication without undue risk to himself or others. He must know how to recruit, incite, train, and direct the operations of agents, saboteurs, resistance groups, and agents provocateur.

b. For Sabotage Training

The saboteur, according to the methods he is to employ, should be skilled in sabotage by resistance, or by destruction, or against personnel, or by coup de main projects. He should be able to reach his objective, perform the act of sabotage effectively, and either avoid detection or effect an escape. He should preferably be able to incite, organize, train, and lead sabotage groups.

c. Morale

The maintenance of high morale is the responsibility of all SO commanders and is especially important because of the hazardous, lonely work of SO operatives. From the time a recruit reports for duty until his service is at an end, building up and holding up his morale is an essential training objective for all officers who have anything to do with the man. SO officers must be personally well-acquainted with each man in their units. Schools and Training Branch officers will inject morale building into their training courses and SO officers will cooperate with the Schools and Training Branch following the progress of their men in the schools. During periods of inactivity or waiting, SO officers will see to it that men are kept occupied with work or diversions directed towards the tasks on which they will be employed and to the maintenance of their morale. Frequent specific checks of the status of morale of each man and each group will be made by responsible SO officers. Senior officers will inspect the units commanded by junior officers to insure that morale is maintained.

SECTION V — SUPPLY AND COMMUNICATION

15. *THE SPECIAL OPERATIONS SUPPLY PROBLEM*

Covering the entire field of sabotage and resistance groups in a number of large theaters of operation means that SO is confronted with a complicated and extensive problem of supply. It will be necessary to obtain thousands of standard items included in the supply tables of the armed forces and in addition many special items necessary to sabotage, underground communication, and resistance groups. Clothing, food, medicines, arms, ammunition, demolition materials, communication equipment, naval equipment, air equipment, money, and other supplies will be necessary to SO activity.

16. *ORGANIZATION FOR SUPPLY*

The OSS Services Officer at field bases or in Washington fills requisitions for supplies, money, and transportation. It will not always be possible to communicate with the Services Officer, especially in active service in the field where supplies may be needed on the spot and immediately. To meet these emergencies SO officers and operatives may be supplied with special funds or through the theater commander authority may be obtained to requisition on vouchers from civilian and other sources. It is essential that all responsible SO officers and operatives have a

SECRET

thorough training in the handling of supplies, transportation, and money.

17. SUPPLIES FOR SABOTEURS AND RESISTANCE GROUPS

One of the greatest obstacles to underground and resistance activity is the difficulty of obtaining needed equipment, and one of the most important functions of SO is to see that the underground and resistance groups receive adequate equipment for effective operations. SO officers and operatives should maintain a continuous survey of the supply requirements of the underground and resistance groups they deal with, report such requirements to the theater or other commander, and make every effort to see that their needs are satisfied.

18. *PAYMENT AND SUBSIDIES*

Special funds are provided for the financial support of underground and resistance personnel. Great care must be exercised in disbursing funds for these purposes as oftentimes an individual activated by money may not be a stable character.

19. *RADIO AND OTHER SIGNAL EQUIPMENT*

The Communications Branch of OSS is the normal source of supply for radio and other signal equipment. All equipment of this type must be obtained through this source.

20. *TRANSPORTATION*

a. Arrangements for transportation of such SO military and civilian personnel as have been requested by the theater commander from the United States to theaters of operation are made through the transportation officer of the theater officer's staff. The necessary passports are secured from the Special Relations Office. Arrangements for overseas shipment of material are made through the Cargo Unit of the Services Branch.

b. Transportation of SO personnel and cargo within theaters is arranged by the Services Officer on the staff of the Strategic Services Officer. When movement of personnel or cargo is required in a place where OSS services officers are not available, arrangements for transportation should be made through nearest appropriate channels of the Army or Navy.

SECTION VI—COORDINATION OF SPECIAL OPERATIONS ACTIVITY WITH THAT OF OTHER OSS BRANCHES AND THE ARMED FORCES AND OTHER AGENCIES OF THE UNITED NATIONS

21. *COOPERATION WITH OTHER OSS BRANCHES*

a. GENERAL

The activities of the branches of OSS are interdependent. SO activities must be correlated with those of intelligence and the other operating branches. SO is part of the OSS team and all of its activities must be planned and executed as part of the OSS program.

b. INTELLIGENCE

SI, X-2, R&A, and FN supply information to SO. Such information will include information from the intelligence services of the armed forces and our allies. SO should obtain its own operational intelligence from the underground and resistance organizations with which they are in contact. Much of the information which SO uncovers will be useful to the intelligence services and others and should be turned over to SI. To avoid duplication of effort and the risk of discovery by the enemy, SO and SI activities in the field will be coordinated for the benefit of both services.

c. MORALE OPERATIONS

The functions of MO an SO will often overlap. Activities of SO may have an effect on the morale of our friends or enemies and SO personnel may be required to assist in MO activities in the field. This will be necessary where MO will not have a field organization, and when MO will train SO personnel to execute MO missions. Sabotage and activities of resistance groups will increase in extent and effectiveness as a resistance spirit is increased by morale operations. MO and SO must work together as each will often be able to aid the other. SO will often require the development of attitudes or states of mind and will request MO to cooperate.

22. COOPERATION WITH SIMILAR AGENCIES OF ALLIED GOVERNMENTS

Our Allies have agencies which in whole or in part parallel the functions of OSS. The governments-in-exile of enemy-occupied countries all have intelligence organizations and are in active communication with the underground and resistance groups in occupied areas. It is the duty of OSS and SO to cooperate with the similar agencies of our Allies. It will often be necessary for SO to be the subordinate teammate of an agency of an Allied government. Every effort must be made to avoid the frictions and misunderstandings which can develop so easily when agencies of Allied governments are working together on the same task.

23. COOPERATION WITH THE ARMED FORCES

The fact that the Strategic Services are under the command of the theater commander is not enough to insure that OSS will most effectively play its part as a member of the military team. It is the responsibility of Strategic Services Officers and special operations officers and operatives to insure that all plans and activities are integrated with the plans of the theater commander. Military plans may call for drastic and sudden changes in the special operations plan and it will be necessary for operatives and officers to conform.

24. COOPERATION WITH GOVERNMENT AGENCIES

Political, diplomatic, and administrative branches of our government and the governments of our Allies participate in the war effort at home and abroad and SO operations must conform to the accepted policies and programs of these agencies. By political and diplomatic activity and through the supply of foods, medicines, and other materials, the government agencies are often in a position to assist in special operations activity. SO must never per-

SECRET

form functions reserved to other government agencies except when duly authorized.

SECTION VII — PLANS AND ORDERS

25. *IMPORTANCE OF PLANS AND ORDERS*

SO activities must conform to the missions laid down in OSS special programs or in approved projects to be incorporated in special programs. Based upon these missions, SO must prepare, in coordination with all branches of OSS, operational plans for the accomplishment of those missions. SO must see to it that SO plans are coordinated with those of other branches. SO personnel and units must always be prepared to act promptly and decisively in furtherance of those plans when an opportunity presents itself. Unless plans are based on accurate information and worked out in exact detail, SO operatives and agents will be working at a great disadvantage. Slipshod planning will result in discovery by the enemy, heavy casualties, and failure. A failure means that SO methods will be revealed to the enemy, putting him on guard, and making it difficult or impossible to succeed after the failure.

26. *ORGANIZATION AND RESPONSIBILITY FOR THE PREPARATION OF PLANS AND ORDERS*

a. The over-all responsibility for OSS planning is stated in Section IV, Provisional Basic Field Manual for Strategic Services.

b. Within the scope of approved Strategic Services programs, the chief of the SO Branch in Washington or at a field base is responsible for the preparation of operational plans and orders covering SO activities. Similarly, the commander or chief of any SO activity in the field is responsible for the preparation of operational plans and orders for the personnel engaged in that activity.

27. *PROCEDURE IN OPERATIONAL PLANNING*

Planning is a continuous process in which all responsible officers participate. It will be the duty of the chief of SO branch or section to develop operational plans covering the missions included in Strategic Services programs. He

SECRET

will also prepare operational plans for activities which the theater commander desires to have accomplished in connection with military operations, and which have not yet been included in OSS special programs. Within the limits of security control a description of such activities will be sent to OSS, Washington, to be included in OSS special programs, which are to be executed within that theater. The process of preparing operational plans and orders will vary widely according to the situation. A plan may consist of a simple verbal recommendation and an order may be an equally simple verbal instruction. Another plan may call for months of detailed preparation and the development of the corresponding orders may likewise entail laborious work. Procedure must never impede effective operation, and when the preparation of formal orders threatens to slow down action, oral orders must be used. The United States War Department Staff Officers Field Manual, FM 101—5, may be consulted with respect to forms for operational orders. The Strategic Services detachments within the theaters are subject to the direction and control of the theater commander and an adherence to military procedure will facilitate OSS work.

SECTION VIII — SABOTAGE TECHNIQUES

28. *DEFINITION*

Special Operations sabotage includes all secret physical subversive activity which destroys or impairs the effectiveness of enemy resources, production, personnel, materiel, and installations.

29. *PLANNING SABOTAGE*

The planning of sabotage will cover a large range of subjects from the most simple act to the highly scientific operation involving inconsiderable original research. Once a sabotage task has been decided upon, careful plans should be prepared for its accomplishment. The enemy will always have a defense against sabotage and no plan can succeed unless this defense is penetrated successfully. Even in the most violent and open sabotage, surprise, deception, and withdrawal are fundamental to planning.

30. *TRAINING OF SABOTEURS*

For all types of sabotage, including the most elementary, the personnel employed should be thoroughly trained in the use of sabotage implements and devices as well as concealment, deception, and withdrawal. For each specific sabotage task individuals or groups should be specially selected, trained, and rehearsed. The details of basic training for sabotage are covered in the courses of the Schools and Training Branch of OSS. For the training of operatives and agents for specific tasks, information and assistance will be obtained from the intelligence services of OSS who will provide information from all other available sources, military, governmental, and civil.

31. *TYPES OF SABOTAGE*

a. INDUSTRIAL SABOTAGE

Industrial sabotage includes attacks on natural resources such as mines, oil wells, and water supply; attacks on processing and handling facilities such as refineries, smelters, factories, and warehouses; public utilities such as electric, telephone, railroad, road, water, and gas systems; and, essential supplies such as forage, foods, and medicines. Physical attacks on management and labor personnel are part of industrial sabotage.

b. MILITARY SABOTAGE

Military sabotage includes attacks on lines of communication, supplies, installations, equipment, materiel, and personnel. Included are roads, railroads, waterways, and their equipment; aircraft, airports, and their installations; radio, telephone, and telegraph systems; food, water, arms, ammunition, medical, and other supplies; key personnel, staffs, sentries, outposts, bridge and other guards.

c. POLITICAL AND PUBLIC SABOTAGE

Political and public physical sabotage covers the liquidation or physical harassment of political and administrative leaders and physical interference with their effectiveness, the demoralization or terrorization of the population by physical means, and physical attacks on collaborationists.

32. *METHODS OF SABOTAGE*

a. Sabotage Applied to Individuals

Includes liquidation, capture, delays, interferences, and physical attacks on personnel.

b. Sabotage by Destruction

Thousands of destructive methods are available including explosions, fires, floods, wrecks, accidents, leaks, breaks, overwork of machinery, maladjustment of machinery, and the adulteration of lubricants, fuels and products.

c. Sabotage by Resistance

Physical resistance by riots and mob action is best conducted by native resistance groups. SO contributes by giving support, supplies, and when necessary, leadership. MO contributes by inciting and instructing resistance groups to acts which impede the enemy's military progress, such as absenteeism, slow-down in production, and other acts of passive resistance and simple sabotage. Sabotage by resistance may result in overlapping functions of MO and SO. Hence, in this field MO and SO must cooperate and coordinate their activities. (See the Provisional Basic Field Manual for Morale Operations.)

d. Coup de Main Projects

Coup de main operations are usually attacks against important targets and are executed by a carefully selected and trained group of SO operatives.

e. Defense Missions

The defense mission is one that is designed to prevent the destruction of installations by the retreating enemy. This includes protection of important bridges and tunnels; wire communications, including wires, transformers, repeater stations; power plants, radio stations, water and sewage systems. It also includes activities to prevent the mining of roads by the enemy, the blowing up of supply dumps, as well as other activities that will prevent the enemy from impeding the progress

SECRET

of the invading forces. Resistance groups, under the guidance of SO operatives, will be the primary agency in the accomplishment of defense missions.

SECTION IX—MISCELLANEOUS SPECIAL OPERATIONS FUNCTIONS

33. *ADDITIONAL FUNCTIONS*

As a member of the OSS—Military Team SO may be called upon to perform a variety of functions in support of the Armed Forces, other branches of OSS and governmental agencies of the United States or its allies.

34. *MORALE OPERATIONS ACTIVITY*

SO may be required to execute field activity for MO. MO activity may include: physical activity for MO effects; the subversion of important individuals; the distribution of subversive pamphlets, posters, or the marking up of slogans; the creation of riots and disturbances; the work of agents provocateur; the spreading of rumors; incitement to resistance; and countering the effects of enemy morale operations.

35. *INTELLIGENCE ACTIVITY*

SI may call upon SO to gather information and to transmit it. X-2 may ask SO operatives to assist in discovering and neutralizing the work of enemy intelligence agents.

36. *ASSISTANCE TO THE ARMED FORCES*

SO may be called upon by theater and other commanders to perform special activities such as to provide guides, interpreters, couriers, and signal men, and to defend or protect installations within the enemy areas. In support of the military plan SO may be required to create diversions with false signals, sabotage, and attacks by resistance groups for the purpose of deceiving the enemy.

37. *DIRECT CONTACT WITH AND SUPPORT OF UNDERGROUND RESISTANCE GROUPS*

SO will maintain liaison with resistance groups; to encourage, instruct, and direct them, and to supply them

SECRET

with munitions, food, medicines, communication equipment, and other materiel.

38. SPECIAL OPERATIONS NOT ASSIGNED TO OTHER GOVERNMENTAL AGENCIES AND NOT UNDER THE DIRECT CONTROL OF THEATER OR AREA COMMANDERS

From neutral areas or in areas not under a military commander, SO may recruit and train personnel or conduct operations in enemy or enemy-occupied countries as directed by Strategic Services in Washington, Chief of the OSS Mission and at field bases. For this type of operation, instructions must be clear and explicit to make sure that SO does not overstep its authority or clash with any other agency, or provoke undesirable diplomatic or political complications. The Chief of the Diplomatic Mission should be advised of such contemplated operations.

SECTION X—THE SELECTION OF SPECIAL OPERATIONS TASKS AND MISSIONS

39. *TYPES OF TASKS*

In sabotage and in contact with and support of resistance groups there is a large field of possible SO tasks, including:

a. ORGANIZATIONAL TASKS—the recruiting of agents, gaining contact with and establishing good relations with such groups, assisting in their training, organization, leadership and supply.

b. OPERATIONAL TASKS

(1) Sabotage of enemy resources, productive facilities, personnel, materiel, and installations, as well as protection of vital installations and equipment required by our own forces and the civilian population.

(2) Miscellaneous special operations tasks in support of the other branches of OSS and the Armed Forces and governmental agencies of the United States and its Allies.

SECRET

40. *SELECTION OF MISSIONS TO COORDINATE WITH THE MILITARY PLAN*

As SO is a member of the OSS—Military Team it is necessary that its activities always be in proper relationship to the military plan of the commander. The status of military activities will have a direct and important bearing on the type of special operations engaged in.

a. DURING A RELATIVELY STATIC OR PREPARATORY PHASE OF MILITARY ACTIVITY

(1) Such a phase may extend over a long period of time during which the opposing forces will be gathering strength or breaking down resistance by bombing from the air and submarine warfare, or maneuvering for strategic advantages on the flanks or by the clearing of lines of communication. During a preparatory phase the activities to be engaged in depend on the situation. However, attacks on military communications, installations, and personnel can be effective during a preparatory phase when the enemy is operating in extremely hostile occupied territory, far from its home base, with limited and vulnerable lines of communication. Under such favorable circumstances, activities of resistance groups can make it extremely costly for the enemy to hold the territory and maintain communications.

(2) Industrial sabotage will reach its greatest effectiveness during a preparatory phase of military activity and the primary objectives should be those facilities whose destruction will cause maximum inconvenience to the enemy. The selection of industries to attack will depend on their relative importance to the war effort and this will depend upon the over-all production position of the enemy. Only a careful and accurate survey of the production picture, industry by industry, will enable SO to determine what objectives to attack and then a full knowledge of manufacturing techniques will be necessary before the best targets can be selected. As a general rule, critical materials and sources of supply, bottlenecks of pro-

SECRET

duction and vital storage and transportation systems should be selected. The foregoing should not preclude the application of general sabotage to anything and everything which may hurt the enemy, if and when included as part of an approved program. These activities should be very carefully coordinated with air intelligence and the air bombing program.

(3) SO may also contribute to an MO program of encouraging slow-downs, mistakes, confusion, demoralization, absenteeism, riots, disturbances, and resistance of all kinds as long as they do not interfere with calculated attacks on the more important objectives.

b. During and Just Preceding a Period of Intensive Military Activity

(1) A period of intensive military activity may include air, land, or sea battles or combined operations; offensives, retreats or sieges; warfare of movement or position; landings or river crossings; and, the campaign may extend over large or small areas of land or water and involve large or small forces. During such a phase SO activity should be concentrated on those missions which will give direct and immediate aid to the armed forces.

(2) Missions may include attacks on enemy personnel, materiel, and communications and they may include defenses of communications and installations which the commander may wish to protect from enemy demolition.

(3) The selection of specific missions will depend on the situation and the military plan. Under one set of circumstances, it may be necessary for SO to concentrate all its efforts on blocking enemy transportation. When the enemy forces are not too strong and are operating in a hostile territory, a general organized resistance on the part of the civilian population may give the greatest help to the military commander. In selecting missions, every possibility should be considered and carefully examined in relation to

SECRET

other possibilities and the military plans before recommendations are made.

41. *AUTHORIZED MISSIONS*

In general, SO activities will fall within the scope of its prescribed functions, as described in pars. 1 and 3, Section I. In the field, these may be modified as the theater commander requires. However, all SO missions must be included in approved programs covering the accomplishment of definite objectives.

42. *TASKS SHOULD BE PRACTICAL*

Unless it is reasonably feasible to accomplish the task assigned with the personnel and equipment available, such SO task should not be undertaken. This does not mean that SO should be unwilling to take risks. SO should always be on the offensive, planning and executing its activities in an aggressive spirit and willing to accept considerable losses and to risk failure.

43. *MISSIONS MUST BE APPROVED BY THE MILITARY COMMANDER*

The responsibility for success of military operations rests with the commander. For security reasons, it will not be possible for SO to be acquainted with all of the military plans. It is essential, therefore, that all SO missions within theaters be acceptable to the theater commander and be approved by him.

44. *CHECK LIST*

In *Appendix "A"* there are summarized in the form of a check list a number of the more important points that may have been presented in this manual. This check list may serve as a brief list of reminders to SO personnel to assist them in the course of their work.

SECRET

APPENDIX "A"
to
SPECIAL OPERATIONS FIELD MANUAL —
STRATEGIC SERVICES
(Provisional)

CHECK LIST

For SO (Washington)

PLANNING

1. *AUTHORITY*

Does the projected activity conform to approved Strategic Services special programs or to additional activities approved by competent authority for inclusion in special programs?

2. *PLANNING IN IMPLEMENTATION OF PROGRAMS*

a. Is planning complete, covering tests as to suitability, feasibility, and practicability?

b. Have provisions been made for:

(1) Coordination of planning with appropriate allied agencies?

(2) Recruiting and training of necessary personnel?

(3) Equipment, supplies, funds, and administrative services?

(4) Adequate communications?

(5) Transportation to the theater?

c. Have SO plans been coordinated with those of other OSS branches to ensure perfect teamwork and to avoid duplication?

d. Have these plans been approved by appropriate authority?

e. Has all pertinent intelligence been forwarded to the field for use in current and further operational plans to be made there?

SECRET

f. Has the field been informed of the steps being taken by the various branches of OSS, Washington, for the implementation of the approved special programs?

g. Have all standing instructions in respect of SO activities been complied with?

SUPPLIES: PROCUREMENT, TRAINING, AND EQUIPPING OF PERSONNEL

3. *SUPPLIES*

a. Has the field been consulted regarding supply requirements for the special programs?

b. Based on that information have lists of supplies and equipment required for the projected activities been prepared and submitted as a requisition to Procurement and Supplies Branch?

c. Has close liaison been maintained all the way with Procurement and Supply to determine:

(1) Availability of supplies and equipment?

(2) Time required to obtain such material?

d. Has the base been notified of what part of the supplies will be sent from Washington?

e. Has branch chief in the field been notified to initiate requests for supplies and equipment as soon as need can be foreseen?

f. Has the field been informed of new special devices and weapons that have become available since plans were made, and have descriptions of their functions and operating details been sent to the field, as well as the quantities available?

g. Has provision been made for adequate funds for the activities under this program?

4. *SUPPLIES OF OSS FUNDS AND SPECIAL EQUIPMENT FOR RESISTANCE GROUPS*

a. Has the field provided detailed information regarding needs of the resistance elements for money, supplies, and equipment?

SECRET

(1) What is available from stocks at the base?

(2) What has to be shipped from the U.S.?

b. Have all needed steps been taken to obtain these materials through Services — Procurement and Supply?

5. *SHIPMENT OF SUPPLIES*

a. Has theater commander approval been received from field for shipment of supplies and equipment?

b. Has field been informed of:

(1) Schedule of shipment of supplies and equipment?

(2) Shortages in the shipment?

6. *PERSONNEL AND EQUIPMENT*

a. Has personnel about to be sent abroad in connection with prospective activities been examined individually for:

(1) Proper training?

(2) Proper inoculations for overseas service?

(3) Regular equipment and special equipment?

(4) Careful security check?

b. Has plausible "cover" been worked out and approved?

7. *TRAINING OF PERSONNEL*

a. Has continuous contact been kept by the SO officers with men in training? Has that contact been maintained in a manner consistent with security?

b. Has special training for the specific assignment been completed satisfactorily?

c. Has the trainee been informed as far as possible consistent with security, of his proposed assignment?

d. Has indoctrination of personnel been completed?

e. Has special emphasis been placed on security throughout the training course?

f. Are you satisfied with the security and discretion of the individual?

SECRET

g. Has the individual been thoroughly coached in his "cover" story?

h. Has provision been made for utilizing this personnel in event of delay in transportation?

8. *REPORTS*

Have you arranged with the field to send you detailed reports of:

a. Operational plans made in the implementation of special programs?

b. Successes or failures in the field in the effort to carry out the missions?

c. Effectiveness of any special devices?

d. Any new methods developed for the use of special devices?

e. Status of personnel—by activities under programs?

f. Cooperation received from pertinent allied organizations?

g. Supply of resistance forces:

    (1) Supplied directly by OSS?

    (2) Supplied directly by the theater commander?

9. *TRANSPORTATION OF PERSONNEL*

a. Has theater commander approval been given to transportation schedules for personnel?

b. Have all the proper documents been prepared and all authorizations received?

c. Has overseas security check been made by OSS, Washington?

d. Has final security check been made?

e. Has final inspection been made of physical condition and equipment of personnel?

f. Has the field been notified giving names, grades of personnel being sent, as well as the numbers that are to follow, if any, to complete the allotment for the projected activity?

SECRET

CHECK LIST
For SO (Theater)

PLANNING

1. *AUTHORITY*

Does the projected activity conform to approved Strategic Services special programs or to additional activities approved by competent authority for inclusion in special programs?

2. *PLANNING IN IMPLEMENTATION OF PROGRAMS*

a. Is operational planning complete, covering tests as to suitability, feasibility, and practicability?

b. Have provisions been made for:

   (1) Coordination of planning with appropriate allied agencies?

   (2) Recruitment and training of necessary additional personnel?

   (3) Equipment, supplies, funds, and administrative services?

   (4) Adequate communications?

   (5) Transportation to, within, and from the area of action?

c. Have SO plans been coordinated with those of other OSS branches to ensure perfect teamwork and to avoid duplication?

d. Has the plan been checked against pertinent intelligence from all sources?

e Have instructions been included in the plan for training of personnel and indoctrination in security and responsibility in the projected activity?

f. Has provision been made in the plan for prompt reports of field personnel to base:

   (1) Information obtained?

   (2) Progress of activities?

(3) Additional assistance required — supplies, funds, equipment, personnel?

g. Has provision been made in the plan for the inclusion in the required biweekly reports on all activities to SO in Washington, of:

(1) Copies of operational plans as soon as security conditions permit?

(2) Effectiveness of any special devices?

(3) New methods developed for the use of special devices?

(4) Status of personnel — by activities under programs?

(5) Cooperation received from pertinent allied organizations?

SUPPLIES: PROCUREMENT, TRAINING, AND EQUIPPING OF PERSONNEL

3. *SUPPLIES*

a. Have requirements for supplies and equipment been carefully worked out?

b. Have arrangements been made with Services to obtain in the theater what is available there from American and allied military supplies?

c. Has Services requisitioned the remaining needs from Procurement and Supplies Branch in Washington?

d. Has the final approved list been checked as to time required to get such material to the field?

e. Have descriptions of functions and operating details of latest OSS weapons been received?

f. Has requisition been made for these weapons?

g. Have required funds been requisitioned?

h. Have steps been taken to obtain required amount of foreign currency?

i. Have arrangements been made for adequate disguise and cover for personnel?

SECRET

4. *SUPPLIES OF OSS FUNDS AND SPECIAL EQUIPMENT FOR RESISTANCE GROUPS*

<u>a</u>. Is a continuing check kept of needs of resistance groups for funds, equipment, and supplies?

<u>b</u>. What is available from stocks at base?

<u>c</u>. What has to be shipped from the United States?

<u>d</u>. Have arrangements been made for a continuous supply service to the resistance groups?

5. *SHIPMENT OF SUPPLIES*

<u>a</u>. Has proper requisition been made for items mentioned in "4" above?

<u>b</u>. Has theater commander approval been forwarded to Washington for shipment of items?

<u>c</u>. Has schedule of shipments been worked out with Washington?

6. *PERSONNEL AND EQUIPMENT*

Has personnel on arrival been examined individually for:

<u>a</u>. Morale;

<u>b</u>. Physical condition;

<u>c</u>. Equipment;

<u>d</u>. Training;

<u>e</u>. Indoctrination;

<u>f</u>. Security?

7. *TRAINING OF PERSONNEL*

For personnel trained at the base, have the following points been checked carefully:

<u>a</u>. Has continuous contact been kept by SO officers with men in training? Has that contact been maintained in a manner consistent with security?

<u>b</u>. Has special training for the specific assignment been completed satisfactorily?

c. Has the trainee been informed as far as possible consistent with security, of his proposed assignment?

d. Is the indoctrination complete?

e. Has special emphasis been placed on security throughout the training course?

f. Are you satisfied with the security and discretion of the individual?

TRANSPORTATION OF PERSONNEL

8. *AUTHORIZATION FOR TRANSPORTATION*

a. Have all the proper documents been prepared consistent with the individual's cover or protection and his proposed activities?

b. Has the individual a supply of money consistent with his cover?

c. Have arrangements for transportation of the individual to destination been worked out with military authorities?

d. Have arrangements been made to insure establishment of the individual's secret communications with the base?

e. Have all measures covering security of individual's departure been taken?

f. Have arrangements been made for the individual's withdrawal in case of necessity or when his task is completed?